ROLE 3

Curt Last

Thank you to the following individuals for helping with this book:

David S. Riseberg
Joshua McKinney

Thank you to the following organizations for their assistance:

Long Beach Veteran's Administration Healthcare System
The Tierney Center for Veteran Services
South Orange County Veteran's Center
Squaw Valley Community of Writers
The Foundation for Art & Healing
Disabled American Veterans
Veterans of Foreign Wars
Sempre Fi Fund
The USO
US Vets

Acknowledgements

A ROOM OF CRIES AND PAIN	The Chiron Review/GWOT
ANGEL	GWOT
BEDCARE ON A DETAINEE	War, Literature & the Arts
CHOW CONVERSATION	Versus and Curses
COAGULATED BLOOD	War, Literature & the Arts
DEATH CARE	The Chiron Review
DOUBLE AMP LIEUTENANT'S WIFE	The Chiron Review
GONE	Veteran's Writing Project
HUMVEE REAR COMPARTMENT	The Chiron Review
INDIFFERENCE	Versus and Curses
MILITARY LIFE AND ART	East Coast Literary Review
MOMENTS WITH A FALLEN SOLDIER	The San Pedro River Review
PRIMARY DRIVER	Versus and Curses
RANDOM OBSERVATION	The Chiron Review
RIPLEY	The San Pedro River Review
TANNING IN AFGHANISTAN	The Chiron Review
WALKING THROUGH LAX	The Chiron Review
WHEN A PAGE COMES IN	The Chiron Review

PREFACE

I joined the United States military for the purpose of rounding out my life experience in order to help fuel my writing. Under the influence of poets and authors with names such as Bishop, Hemingway, Orwell, and Melville, I chose travel and worldly experience as a key part of my growth. Life experience tends to create a deep reservoir for the Muse to siphon material. Joining at a later age as a well-educated, well-read, and well-travelled individual, I volunteered for deployment in hopes of doing meaningful work as a Reservist sailor. I went into the warzone with a cheap camera, a journal, and books shipped from home, knowing this was an experience I would write about. The journal would end up being my strongest memory device, and possibly the one way I maintained sanity with all of the things going on around me at the Role 3.

*   *   *

I've been asked, "Why not write this as a collection of short stories, or a memoir?" Simply, poetry is the form my Muse tends to emerge from, and the medium which suits my want to express my ideas and experiences. It is where I am most comfortable, and I enjoy the great challenge of remaining 'poetic' while having a predominantly narrative approach of interwoven poems with some stand-alone pieces and minor experimentations with form and thought construct.

I prefer the narrative form as I feel I have ideas and experiences that need to be projected as clearly as possible, and a lyrical or objective approach may dilute them. I'm not interested so much in appealing to the cognoscenti or literati. They may offer a reality check in regards to the poetry within the work, but ultimately I prefer an appeal to the average reader, if one can use such a classification. I wish to garner interest in what I am trying to convey in someone who would passively pick up this book. I believe the narrative form brings in a wider audience.

*   *   *

Though the lyrical war poems of Yusef Komunyakaa and other war poets are beautiful, I want to spell things out in a concrete manner, keeping everything simple and close to the language we used as military medical personnel in a warzone. The experience has to be projected as clearly as possible in all of the ugly details. I feel the need to affect a passive audience and involve them in the understanding of war and violent death. As part of that process, everything in this collection has been broken down to its most simple form—from the book title to the repetition of language and use of terms. For example, "flight line" is used often, and refers to the actual physical airstrip as well as an abbreviated term for the 9-Line crew's leader:

Example 1: We arrived on the flight line ready to pick up wounded.

Example 2: Flight line directed us toward the Blackhawk's cabin door.
Flight line is also interchangeable with Foxtrot, which was the designation for the runway portion of the airstrip at Kandahar Airfield. Similar usage occurs with 9-Lines as the actual 9-Line call or the wounded individuals on the call. "Trauma bay" includes the actual trauma area, the outside area where we unloaded wounded to be transferred inside, as well as individual bays with single beds where wounded were laid out for examination and treatment. Context should clue the reader in without causing confusion.

*   *   *

When the reality of an experience outweighs the attempts of art, one ends up with a collection of poems such as that which I have created. I've been told that it feels like the art is trying to escape from the military philosophy and way of life. I believe the art has succeeded in doing so and is attempting to coexist with the military experience. The concrete, often simple and repetitive use of language lends itself to the accuracy of the experience. I would find it insincere to try to add extra colors and brush strokes to the images in my head. I feel the work does justice to the experience, to those who were over there with me, and to the unfortunate individuals we treated. Role 3 was known as, "the best part of the worst day in someone's life." Now Role 3 is part of any willing reader's life.

CONTENTS

HOME: 01SEP2011 - PRESENT

INTRODUCTION

I deployed to Afghanistan like many other Navy medical personnel—as a volunteer. Our Role 3 rotation was a collection of Navy officers and enlisted, both active duty and Reservist. We were brought together from Navy hospitals on Guam, Okinawa, Hawaii, and all parts of the mainland. Some of us processed through Port Hueneme, others processed through San Diego or Portsmouth, Virginia. We were then linked up into one working medical unit and sent to Fort Dix, New Jersey, training in deep snows during the month of January. We were then sent on our way through an airbase in Germany, then a week in Kuwait processing for our arrival into Kandahar, Afghanistan.

As a general duty corpsman, or quad zero (0000), I didn't have the distinction of running x-rays, ct scans or lab work, putting together medications and IV fluids, or working in the surgical suites (though at times I had the fortune to visit friends in these departments, or was given the opportunity to watch a surgical procedure and participate in a minor way). I started the tour on the Urgent Care crew running day shifts, where we functioned as clinical corpsmen assisting US Army and Afghan walk-ins. We were also the first responders, responsible for getting out to the airstrip and bringing in dead and wounded.

After two months, my Urgent Care day crew switched to nights. At 1800 Urgent Care officially shut down, so I would be released to work in other wards when not responding to 9-Lines. I worked predominantly in the ICU with the nursing officers there, who shared their knowledge and allowed me to assist in some of the more involved processes and procedures. A good bulk of the work herein was created from events that took place during this time period.

I was finally transferred over to the Inpatient Ward for my last 2 months of deployment in order to allow incoming Urgent Care corpsman training space. The latter portion of patient poems covers this ward, which was broken down into Ward 1—the Children's Ward, Ward 2—the Coalition Forces Ward (America and its allies), and Ward 3—the Local National Ward. There were also two additional rooms with two beds in each for detainees.

*   *   *

On August 24th, at 1530 Kandahar Local Time, the wheels of our C-130 left the ground. The trip back home involved some mild decompression—with a week-long 'R&R' in Kuwait on an American base—where I ducked out in Starbucks, reading a novel on the history of Mexico. We were flown through Germany, then to the East Coast, where we finally broke off back into our original indoc groups. I demobbed at Port Hueneme and spent all of 2 days processing out.

Labor Day Weekend Friday, September 2nd, 2011, I was wearing my civilian clothes and a beanie (not the one I was "ordered" to remove in the poem "Sangfroid"), riding a crowded Amtrak train back to Orange County with a noisy vacation crowd. A week earlier I had been in ACUs in Afghanistan.
It was a strange homecoming. No parades, no hanging out with the family or friends. I hadn't seen my girlfriend since December 2010, when we met in Hawaii right before my deployment. I wouldn't see her again until I flew out to Japan at the end of September 2011. This distance and physical detachment from everyone and thing made me realize how hard-boiled I truly became. Being over there thickened my shell (while in some ways thinning it out). In reflection, I don't think working on this book has helped my perspective, as I always felt 'in it' and observant and reflective even as events occurred. However, it has forced the Muse see things in a different light.

Hopefully this work will earn a receptive reading audience who is truly interested in gaining some understanding of what our people continue to go through in warzones. I can only offer what I have personally lived through and observed.

Curt Last
Huntington Beach, California
September 2015

When the grave lies open before us, let's not forget,
but make it our business to record the worst
of the human viciousness we've seen
without changing one word.

—Ferdinand de Celine, *Journey to the End of the Night*

KANDAHAR AIRFIELD (KAF), AFGHANISTAN
19FEB - 24AUG2011

IN

First two weeks in a dusty tent,
through all of the nights awakened by jets
and helicopters somewhere far off—
rigid mountain chains echo the drumming.

Base mini-guns and cannons blaze off—
killing stillness—all just weapons checks,
or is our perimeter threatened?

Fine dust shrouds every surface—
breathed in while doing anything outside—
vehicles and the wind throwing it up
from base asphalt and gravel roadways.

Concrete slabs positioned for road blocks
and bomb blast protection are everywhere—
drab and square—highlighted by barbwire.

Reek of port-a-potties' chemicals mixed with
old piss and the warming weather brings
a mass of flies—a hint of death to come with
the Taliban Spring Offensive.

THE BOARDWALK

Imagine the filth of Tijuana meets the multi-ethnicity of a busy city,
all in a military base setting within a warzone, and you have KAF

—journal entry

Wandering to the Boardwalk in the afternoon, wood construction as a Wild
West town built into a defensive square: TGIFriday's, a smoothie shop, KFC,
coffeeshops, a dry cleaner, a beauty salon—manis and pedis done upstairs,
local merchants sell junk wood boxes and bejeweled metal animals imported
from somewhere, labeled 'local hand made' and priced for a seller's bazaar;

and the troops—US Army, Marines, Airmen, Sailors, Slavs, Brits, Aussies,
French, Dutch, Germans—geared up with pistols at hips or in shoulder holsters,
heavy weapons slung over shoulders or dangling across crotch Jimmy Page Les Paul
straddling low—it seems we can get along, because no one has drawn;

and the civilians—Nepalese, Indian, Paki, aging American and Brit contractors on
their last leg of good pay in a shit war zone, pretty Chinese nail salon girls (among
other jobs), a money rainbow of force and culture, a NATO war project, the feel
of the Star Wars cantina at times. Not quite "a wretched hive of scum,"
but are we not the Imperial forces, Obi Wan? Does a Sith mind trick deceive us?

WHEN A PAGE COMES IN

It may be slow in Urgent Care,
ailing soldiers filling up bays,
each corpsman assigned one—
responsible for one patient at a time,
while a check-in corpsman adds patients
in AHLTA, guides them back, scribes stats
on a multi-ink-coded board,
as their corpsman works them up,
gives report to the nurse or doctor
or nurse practitioner on duty.

I may have a needle in a hub
to get blood for a WBC
on a suspected appendicitis
or conducting a gonorrhea swipe—
not thrilled to hold a dick not my own.
These Army guys might be out there
for weeks—pain close to bursting
with festering wound or growth that
dirt and duty exacerbate,
then they're at Role 3 Urgent Care.

Suddenly pagers sing through hallways and bays.
Our charge nurse—always on top of things
having consulted trauma docs
and confirmed details—as traumas teams
often receive TOC's information
before Urgent Care's first responders.
CDR Caballero relays status
to the day's assigned flight line leader
who sets up a response team and
logistics after checking with TOC.

I may have a pager, or hear one,
put down equipment, check call details,
inform my patient I have to leave,
notify my nurse, who already knows—
chaos of energy echoed by walls—
pagers chirping and corpsmen running,
grabbing coms, cranials and crew mates.
If not on EVOC, I ask flight line
if an extra hand is needed.
Always rushed—sometimes "no," sometimes "yes,"

Then it's the quick dash to the wood cubbies,
grab my DCU blouse and cranial—
clothing and helmet worn for protection
while on the airstrip flight line deck
under the rotor wash of Blackhawk blades.
EVOC1 holds a walkie talkie,
EVOC3 as well—secondary Humvee
has to keep track of any status change:
EVOC1 to TOC, TOC to control tower,
control tower to incoming flight crew.

We burst through doors—run past trauma bay,
grabbing latex gloves for BSI,
pile into two Humvee ambulances.
Engines start, headlights and blinkers glow,
chock block stops are pulled out from front tires,
the static squelch of recognition
as EVOC1 verifies team outbound,
primary vehicle followed by
secondary, makes way to landing zone Kilo.
Engines idle, crews sit and wait.

Stacked up on a white line off of Foxtrot,
just off the runway flight line,
we scan the whole blue distance for helos.
Birds come into view, fast approaching,
flight line leader notifies TOC
of visual—"TOC, eyes on bird,"
as blockers—Air Force flight line security—
pull out on opposite ends of the runway
to allow us zero interference,
engaged in saving the lives delivered us.

Our two Humvees roll out on the airstrip,
park and idle parallel to each other,
litter crews jump out of rear compartments—
hold onto any part of Humvees
as first flush of helo wash whips up
a maelstrom of Kandahar moon dust,
settles to a consistent heat windstorm,
Litter crew, crouching, moves toward the hatch,
drivers and flight line intently watch
as each corpsman performs an assigned role.

A SOLDIER SUCCUMBS

his eyes were a distance, somewhere alive—here
or the other side; his mind may have been clear.
maybe he was moving toward the face of his god,
thinking of a loved one, or only wondering when
the cold shock would end, if thought could muster.

PROCEDURE

Two of our sailors
pace out
to the flag pole,
lower the ensign
half-mast,
an internal
email passes
through the base,
then all flags
lower
half-mast.
The internet is killed—
someone
may drop
opsec
online
before the family
can be notified.

THE BODY TAKES SO MUCH

The Afghan soldier
had his face blown off—
a bloody red skull
of exposed maxilla,
mandible and teeth,
and the x-ray tech
later in disbelief
says, "Man, you see how much
the human body can take,"
but his words aren't inspiring
—no one comes back
from that kind of wound
exposure—the military
hospital in Bethesda
and the best plastic
surgeon in Beverly Hills
couldn't keep him going
long enough to give him
a Hollywood face.

PAGER DESIGNATIONS

EVOC1

Flight line leader
sits in the back compartment
of the primary vehicle,
initiates coms with TOC
as the two Humvees roll
out to the flight line
to await the incoming,
views the horizon for helos,
standing off the runway
until black dots appear.
"TOC, eyes on bird,"
he squelches over coms
to update status.
Humvees cruise out, park,
litter crews jump out,
rush under
whipping blades,
carry wounded—
a lift of the litter—delivered
to flight line leader's waiting grasp,
who pulls and slides wounded
along metal rack,
secures O2 mask,
pumps chest
with the help, or derogatory comments,
of the Army flight medic
who boards to assist
what was their patient
back in the helo,
until they ended up
in the back
of one of our Humvees
and in flight line's hands.

EVOC2

Primary driver, first Humvee
out of the lot,   stacks up
on the white line   just off
the airstrip flight line—

the Air Force CASEVAC
smoke deck   right next to
the white line

          (it's surreal to watch
          off-duties casually smoke
          as a life is in balance).

EVOC2 will takethe first patient,
hopefully the one   in the worst condition,

two   if more than one Alpha—
designation for   the most traumatically wounded.

The drive back   calculated,
slow—   not to jar the wounded,
only to get them to trauma doctors
inside the hospital.

With the patients admitted
and paperwork turned over,
primary driver takes the pj back

He boards   helo lifts,
hovers, spins,   and heads back out.

EVOC3

Secondary vehicle either crawls
back to trauma bay
with the litter crew
to assist with movement
of seriously wounded,
or loads up Bravos and Charlies,
or carries escorts
accompanying the wounded—
a buddy with a soldier's gear,
a relative of a wounded Afghan.
On bad calls—or worse—
EVOC3 will standby
for additional Alphas and Bravos.
Rarely do both Humvees
go back out to flight line for more,
though sometimes the carnage
leaves the litter crew running
back to trauma bay,
no room aboard, but for wounded.

EVOC4

The lead litter bearer
will always hold
the left head of a litter,
to the patient's right shoulder,
directing the rest
of the four-person litter crew
to the helo—
guiding them
in receiving the patient—
delivered in thermal wrap,
torn clothes and blood.
The lead litter bearer
guides the carry
to the Humvee's door-less
rear compartment,

ensures the patient is lifted headfirst
into the receiving hands
of the flight line leader,
who radios TOC
our inbound status.

PRIMARY DRIVER

I drive the primary vehicle, slowly cruise out, watching speed,
and stack up right off the runway, then wait until a visual
is established on inbound MEDEVAC and armed escort. Some
are salsa dancing, others joking—most silent and intent.

Black dots in the distance—they grow—our two Blackhawks,
I pull onto flight line after our blockers roll out on to the airstrip
to our North and South—Air Force-manned Humvees that ensure
no incoming or outgoing aircraft will interfere with our operation.

I peel in a J shape to a point just outside the big landing X of KILO,
just enough to avoid the helo rotors by 20 meters or so, but close
enough to spare the litter crew the burden of extra steps
with a heavy load and additional seconds wasted carrying too far.

The litter crews jump out, grab sides, brace for the helos' wash,
myself and the other driver sit safely in our vehicles, goggles on
to shield eyes from dust, though I keep my helmet on—in case
I'm needed—anything which may call for a change in protocol.

Peering into my side mirror, the litter crew approach the helo,
wounded pulled out, legs first, walked straight out from Blackhawk's
hatch—avoiding the downward spin of blades—turned and spun toward
Humvee so patient goes in head first, the lift of faith, of patient in air.

I pull my body to the right, look inside the rear of my vehicle.
I turn and observe the flight line leader grab the top legs of the litter,
pull the patient up, hobble awkwardly backward, as the rear legs
of the litter slide on rails, he gets down and applies O2 mask.

The pj jumps in, starts debriefing him, flight line leader goes to work,
gives a thumbs up to me—I pull off the runway, a few small turns,
honk at any traffic or pedestrians, pull up to trauma bay, brake pedal,
apply the emergency brake and put it in park as my engine idles.

The patient is unloaded outside trauma bay, placed on a rickshaw,
secured by clamps, as the Master at Arms searches through thermal wrap,
clothing, between body and stretcher making sure there isn't any loose
ordinance that can roll out in our trauma bay and kill one of our teams.

Thumbs up from the MA, two trauma corpsmen waiting inside run out,
kick out the rickshaw legs and roll the patient into the trauma room.
I pull forward, back up with help from a ground guide, over crackling gravel
and park my vehicle. My official duty is complete. I grab my cranial, head back.

TRUE GORE

March 6th and I'm secondary driver when we get the call for an Alpha.
Two weeks in country, the strain of seeing an Alpha blow up on our pagers
shows, while carnage numbs most to goof and play while on standby.

Middle of the afternoon on a hot, clear day, we stack up on standby
off the runway as a MEDEVAC tears in with its gunship escort, Humvees
whip out toward the white 'X' at landing zone Kilo, stop, and set up.

Landing in dust and debris, helo blades whip—air vibrates, Humvees rock,
litter crews crouch, cling to side walls and bumpers, all settles, the four turn
and head toward helo, crouch walking out underneath churning blades.

The sliding door rolls open, the pj pumps his adrenalized arms
to indicate move the *fuck forward now*, the litter crew scurries
in response, views and assesses the patient's layout and needs.

Already on a stretcher, a shadowed body—head and torso only—
leg crew pull, the head team follow up as pj releases, continues to do
whatever necessary as they're moving with cautious, controlled speed.

Coming up toward them, watchful of carry, stunned by what I see
bouncing on the stretcher—a young soldier completely torn in half,
a wall of red jelly insides divides his body from the world surrounding it.

Missing his lower torso, the body dances unnaturally
as the litter crew moves as ghosts with him. I scan—no guts, no bone
nor movie blood—a clean, medial view of a man torn by IED.

Oh shit, he's going out! I yell in my mind as I silently watch him lifted
into our first Humvee, jump back behind the wheel of the second
Humvee, check that litter crew are on board, and go.

Primary pulls into trauma bay's shadow, litter crew jumps out of secondary,
idling now, I'm out and to the side, out of the way, enough corpsmen
helping unload and secure the dying soldier to rickshaw.

I get a good look at him, a view unreal, something I needed to do
in order to clarify what my mind thought it saw, if not to ascertain
the young soldier, to see if I could stomach real, true gore.

His skin is a sallow yellow—blood-drained, head back, eyes to sky, is life still
in them? I imagine it to be, look into them meters from his form, I cannot make
sense—there seems to be something or someone there.

Trauma corpsmen continue compressions, he's jaw agape on the rickshaw litter,
still standing by, off to the side, I yell, "Fight! Fight!" in hopes that he can hear,
in hopes that there's still something left in him, if anything.

My roommate, MA1, has already checked him for any loose ordinance—
grenades, bullets—anything that could pose a serious problem
once he's been wheeled into the trauma bay, lest more carnage ensue.

MA1 calls him "clear," a nurse steps out, the trauma corpsmen wheel
the soldier in, the nurse by his side through the double doors,

now in the full care of a Role 3 doc, yet within minutes he's pronounced dead.

I drive the pj back to the helo—having been with the soldier the whole time,
trauma care in the helo, the Humvee, turning over paperwork in Role 3,
watching his patient pronounced dead.

Now, sitting in the back of my Humvee, I don't bother him—
just making the turns, eyeing the landing zone, shooting a curve to the helo,
idle and observe him exit, board helo, and the bird is gone.

Back inside trauma bay, in formation—I'm part of the procession saluting
the removal of the soldier's body, ceremoniously draped with Stars and Stripes,
to a white mortuary affairs truck smoking and idling outside Role 3.

ANGEL

Another day, and a 9-Line for an Alpha comes in.
So we're out on the flight line waiting...waiting...
until we're out in the daylight heat for an hour
and our Alpha is now an Angel—call sign
for deceased coalition forces—which means
he could be English, Australian, Slovakian
or American, and we feel a different tension,
though not as immediate...
but much more visceral, bringing all of our hearts
down as those two Blackhawks grow from spots
in the thin blue distance from the Southeast.

The body bag is delivered to the primary vehicle
along with an escort in ACUs, as the litter crew
jumps into the back of the secondary, which follows.
I fall out the back and there's the blue canvassed Angel;
on the skid opposite him his buddy sits, head down;
I approach the back ladder and reach out to him,
put one hand under his left arm pit, the other
around his waist, down the ladder, bearing him,
covered in Kandahar dust, bloody mouth and nose,
and he erupts in crying hysterics as I grimace while
holding his weight, guiding him into the trauma bay.

A psych tech is waiting and I let go, keep walking
back to Urgent Care, ditch my gloves in a can,
keep walking, take off my cranial, keep walking, say
a prayer, the little a Buddhist can give to a kid who couldn't
have been more than 19 years old, dealing with the sight
of his friend in a body bag. It's one of many images
which will always stay with me, if not for the suffering,
then for a scene which war movies mock with actors
who cannot have the level of honor, the understanding
of horror, or the quality of truth that art continually,
aggressively pleads a susceptible audiences to believe.

I do not believe in the term 'Angel' for it limits suffering
and places it in the hands of myths and gods.

RANDOM OBSERVATION

Being in a war zone forces you
to see the poetic beauty inherent
in the Twenty-third Psalm.

ROLODEX

Two weeks in:

A burned soldier

A few filled body bags

waiting for the doctor's signature

Blood-saturated, gauze-faced intubations

A face without flesh

A human torn in half.

It's building in my head—

not intentionally

but it's happening.

THE HUMVEE REAR COMPARTMENT

HN Aguillera's eyes start to mist
wiping up blood and dust
from the rear compartment
of my Humvee. I tell him "I got it."
Blood doesn't do much to me.

Gore may make me reel
at times, but the blood is fine,
in small amounts, just the right
amounts where the copper
won't overcome me. Aguillera,

as fresh to this as me, is half
my age, and I wouldn't want to be
23 and seeing all of the things
that he sees. My eyes are a departure
from myself, they must be,

as we're weeks deep into this,
we've seen enough to outweigh
years of trauma in an ER back
home, that first triple amputee
bouncing oddly on the litter...

thank God he was an Afghan
and not United States Army.
Callous as it is, it feels better
to not see our boys like that,
and better yet, that it isn't me.

THE LIMB

It raised hairs like day-old meat
decaying in warm air in a *carniceria*
I curiously strolled through long
ago on a vacation in Mazatlan.

The feel of it was soft, mush—
a sensation one could imagine
if having ever held in their hand
a large, loose, bone-in steak.

No longer enervated by owner,
still, under his ownership, only
government property to be claimed,
still, just meat—dead muscle.

Two staff sergeants arrived by night,
right after a 9-Line that included
a mutilated, clinging-to-life soldier's
duffle bag with everything he owned.

Under the overhang of trauma bay,
under false light and the dying day,
illuminating a story, personal items
and what these men were recovering,

I pulled from green canvas duffle bag,
told them it was all right, I would take
the contents out—I felt obliged to it—
I had accepted it from the helo crew,

slung it over my back, into a Humvee,
possessing it, awaiting their arrival—
to recover his NVGs, reusable gear,
though blood stains-soaked duffle bag,

still, upright, stiffly filled with his life,
I pulled out the limb, ACU legging
and combat boot, wrapped in plastic—
jagged bone puncturing through.

We observed it with a professional
strangeness, my left hand gripping
mushy calf muscle, and one of them
relieved me as if a weapons turnover.

They trodded away from trauma bay,
delivered it to mortuary affairs
for proper handling, and—I'm assuming,
a return to a barely living soldier.

KNOWING

I glance away
from the call,
the 9-Lines lay
safely within
trauma bays.

A large group of
young corpsmen
crowd the line
as teams
methodically treat
broken humans.

Transfixed
like commuters slowing
at a fatal car accident,
all seem caught
under white light;

and if I'm in this room
any longer
I allow nightmares
to grow
from crushing vines
as black bougainvillea.

9-LINES

JOURNAL ENTRY:

Last two days at work were pretty busy. Monday we had eight 9-Lines, yesterday four—but two were back-to-back—the first MEDEVAC and escort had to move up Foxtrot to let the next incoming land. We had 3 Army hit by an IED. None of them looked too bad—no missing limbs, no gaping wounds.

RIPLEY

Ripley—she's the one enlisted person in Urgent Care
who seems to understand the ways in which I don't know shit,
but still wants to help me learn, instead of talking down
to me as if I was a prematurely-born retarded monkey.

A reservist who works at a clinic one weekend a month
at a Navy Operational Support Center, or NOSC,
I draw blood for HIV and Sickle Cell tests, needle arms
for immunizations and TB testing and run vitals for PHAs.

So what do I know about military data entry systems,
proper protocols for patient care, simple processes and names
for medical supplies that these young bone-head kids use
on a daily basis at the Naval Hospitals they work at?

But I'm an idiot because I've never had the opportunity
to learn slowly and thoroughly, or at least through time
and experience such as already mentioned bone-heads.
Worst part is... some of them have rank on me.

HN Ripley is a rank below me, yet hospitable, well-spoken,
and patient—beyond age and rank—which is why I like her.
Not in that way. She is beautiful and intelligent, but married,
too well-adjusted, and much too young for an aging man.

I call her Ripley after a young Sigourney Weaver in *Aliens*.
Confident and independent, she would take on an alien queen
to save a child entrusted to her, yet warm and kind—her rhythm,
that winning smile that slides through the error in my ways.

INSIDE IPAN'S WORKSHOP

HM2 Ipan is one of the few people I like here.
He's married and faithful—something I've found
rare in my short, but observant, military experience.
A professional in what he does—an ortho tech—
he lets me assist with the basics so I may learn.
He places casts on broken limbs, takes care of ex-fixes,
analyzes soldiers for fractures and sprains,
and places and removes sutures as designated.

He allows me to remove the multiple stitches
from the right hand of some Nepalese kid
who cut off two-thirds of his left hand's pinkie
and carved a gash deep into his palm
while prepping lunch meat at a chow hall—
youth on a meat cutter, his hand slipped.
Now over a week later, the Frankenstein hand
is ready to have thread pulled from it.

The process takes over an hour, fifty stitches
laid through the center of his palm, over the ridge
of pinkie stump, woven down the tear to the side
of his hand. He flinches and whimpers throughout,
as I slowly remove black thread from deep in flesh,
coagulated blood a resin—I break clear, red paste,
hold the scissors steady as he fidgets and jerks away.
Flaking thread is pulled from his sealed wound.

M9 GLO-BELT ID DOG TAGS

I always carry my
M9
glo-belt
CAC card
dog tags

the dog tags are in case
I get blown up, or at the least
to make me feel very military-like.

the ID I can't eat, get on a computer,
or drive on base without.

the glo-belt looks ridiculous,
but will save my ass at night—
pedestrians are hard as hell to see
in the darkness and distraction
of shadows, dust and loud noises.

my M9 has become light
up against my right axillary,
the two magazines on my left
seem to hold more weight.

maybe my weapon
is escaping
from the physical
into a metaphysical conceit.

MILITARY LIFE AND ART

I develop habits to feel less militaristic:

writing poetry in green Memoranda pads
supplied for military notes and records
that need to be remembered or passed on

reading literature—books along my desk
at NATO barracks—Conrad, Levis, Fante,
rock yard stones as functional book ends;

though most literature is military-related:

The Peloponnesian War—Kagan
In the Graveyard of Empires—Jones
Imperial Grunts—Kaplan

on my wall: *Reader's Digest* articles on
Poppy Wars and IED'd Marines missing limbs,
a USO calendar, a map of Afghanistan

my Haji bazaar dvd list: *Platoon, Three Kings,
Jarhead, Full Metal Jacket, The Hurt Locker,*
an MMA collection to last an entire deployment

all photos are of the people here, ergo military,
and this dusty base and that sterile hospital,
journaling related to this warzone day-to-day

Art and military life dance, exchange places,
become like one and other—become the other.

THIRTY ROUNDS

I carry a Beretta 9mm,
2 magazines,
15 rounds in each.

I never think of using them.
Who would I use them on?

If anything,

I think about not using them,
but have to carry them as part of my obligation—
a general order.

There are many young junior enlisted
who often seem to wish to load up.

Hopefully we never have to use any of them.

ESCORTS

Escorts fly in with family or friends,
or enter on foot through base security.
I assist PAD's driver in transporting
guests of wounded Afghan civilians.

I check my 9mm out of the armory,
ready to escort them off of KAF.
But who is to say who they are,
and who is truly dangerous here?

Into the hospital's white Nissan van—
usually it carries trauma teams to calls,
or supplies a ride for lazy corpsmen
going back to NATO barracks from duty.

Then there are the hospital's guests—
some released early for suspected TB,
others walking cured, or today—a turbaned
escort, malicious looks and ancient skin.

He sits in the bench behind the driver,
and I sit directly behind him—no allowing
any perspective of what I'm doing,
I like to control the situation—9mm aware.

Every ride I sense the situation's joke,
sending off a hospital guest with guns,
making sure they are beyond the posts,
concrete blast walls and concerto wire.

We walk through numerous turnstiles
at ECP2, groves of barbed wire, chain-link
fence clearings let us know what's ahead,
as we pass security boxes and lines of locals.

Driver and I wave goodbye to our guest,
I'm relieved to see him going, as we pass
through an awkward purgatory of safety,
my senses bearing, caution guiding me.

Looking over to my left, no one is close
to me, numerous Afghan males wait
in shade, a simple wood waiting area
hides them from their own mid-day sun.

Eyes of wolves scavenge into my own.
Do they search for weakness or mock
from deep beneath reflective surfaces
of their own obsidian mysteries?

ROCKET ATTACK

It was a nice night for a stroll,
the moon above freed from dust,
while rockets fell...somewhere,

I wandered by the coffee shop line,
people still waiting for cool drinks,
while rockets fell...somewhere.

Strangely, Poo Pond's smell
was not there, as the hot night
offered no breeze for its air.

In an unchanging furnace at 2300,
reflecting on all of the rocket alarms,
the proper British woman's voice:

ROCKET ATTACK...
ROCKET ATTACK...

Over the base speakers that night,
as so many other nights previous,
down memory lane with her I fled.

My very first rocket attack I was
eating in Cambridge DFAC
as everyone dropped to the ground,

and I followed, though I just wanted
to eat my beans and toast—the CRUMP
of metal crushing, and the WHUMP—

a slight sensation of pressure moving air
like heat moved by wind over a still lake,
as if Poo Pond could be such a place.

There was the night at the National DFAC
where I strolled out, pressed up against a wall
(in case of a good blast), flattened and waited,

but took instead to sneak back to my room—
NATO barracks purportedly rocket-proof,
I assumed it should be better than any DFAC

where a month before we arrived in country
a rocket took one life and left seventeen
wounded strewn across tables and chairs.

Then at Tim Horton's on a Sunday
afternoon—a day off—a book and journal,
a bagel and a cup of java. A minute later

a concrete shelter—with my books,
coffee, and a bunch of Canadians.
Rockets may bring us together, but

most often feel like an earthquake
in California, a pain in the ass
getting in the way of life, in this way

of life.

DETAINEE

The term is vague—as ridiculous
as "good guys," "bad guys," or even
"us" and "them"—and I don't think many
enlisted read Deconstructionist
Theory, as the poor Afghan could be
"at the wrong place at the wrong time";

but I've found when talking to Army guys
guarding wounded suspects in ICU
or the Inpatient Ward that MOST
were caught planting an IED
or in the back cab of a pickup truck
stacked up with a bunch of armed men;

though when I ask guards of detainees
they sometimes don't even know the story
behind the detainment, they were simply picked
to do a boring (and potentially deadly) task—
sit in our hospital, trading off sleep
and head breaks with a buddy.

Often they're from the unit that caught the detainee,
sometimes from an MP unit sent in as handlers.
All desire to return to their buddies in the field,
to be around dangerous familiarity and friends,
to be away from a Haji they would prefer to kill.
I consider their hate, understand their hate

but cannot have their hate—I'm not out there.
So I never shout or laugh at a *Haji,*
and the same ones who have spat at guards
and other corpsmen have never spat at me—
I don't know why, as I restrain my temper,
consider the story behind their detainment.

Guards may hope I mistreat the patient,
though I've never seen a detainee spit, throw feces,
or even curse at anyone—quite often,
they are relaxed, indignant yet resigned.
If I judge a captive, I like to know the circumstances,
what or why they were fleeing when captured.

"ALLAH!! ALLAH!!"

In HM2 Ipan's workshop—'the orthotics room,'
at 30 degrees on an exam table in the bay
closest to the doorway, a short, fat local
in traditional garb and cap yells out to his god,
as the ortho nurse tries to take the ex-fix screws
out of his right upper arm and shoulder area,
its long stainless pins sunk deep into his flesh,
while he cries out one word in short-burst groupings:

"ALLAH!! ALLAH!!" "ALLAH!! ALLAH!!" "ALLAH!! ALLAH!!"

Voice trembling, pathetic, a complete lack of control
in the face of pain—a common trait of the local men—
they tend to cry out like little boys, no exaggeration.
The nurse gives the patient a healthy dose of pain meds
that don't seem to work. Reliance on opiates said to be
common among locals, lowering their pain threshold,
so when real pain occurs, a blocker does nothing for them.
The ortho doc is called—and he's probably pissed
since he's busy with a hell of a lot of surgeries; They move
this whining adult over to one of the trauma bays.

I don't go over with them, but I'm told that at the sight
of the needle the doctor pulls out to give him a higher dose
he flails, almost hitting the doctor's hand
with the needle in it—pissing the doc off even further—
since if he pokes himself, even with a clean needle,
there's a shitload of paperwork and shots for him,
so he loses his cool and yells at the fat baby.
What makes it even more of a pisser is this guy is a detainee.

THE INCUBATION LAB

I'm sitting here in the lab
incubating,

wondering about the
other side—
home.

Considering here
and what it may be doing
to me,
what it might do
to me,

I can only let go,
wait for home,
the same way I surrendered—
waited for here,

something one just can't
put the mind around.

CODY

He's a bilateral amputee—both legs gone
above the knees (those gone as well, logically).

The manic female nurse simply orders me
"Get two other corpsmen to help you change his dressings."

HN Aguillera, HN Stevens, and myself, start work
on the Army Sergeant who almost finished his last patrol.

He's muscular—a six pack, shaved head, and tan—
good looking enough to where I'm sure his wife

will overlook the missing legs—maybe even find it
strangely attractive—mutilation carved into perfection.

We cut away the soaked-through ace wraps and divide
the work—Stevens on the left leg, I on the right,

working trauma sheers through blood-drenched gauze
that protects his stumps, bright red magnifies the white.

Stevens brings the doctor in, after we've pondered
removing all of the bloodied gauze and starting fresh.

"There isn't much bleeding, but he still has two open areas
at the stumps to allow seepage. They only need reinforcement."

We ask Cody—as he's requested we call him by his first name—
to lift his stumps—a great effort, as Stevens and I go about

putting fresh 3-inch gauze over what has been bled-through—
over, under and around the stumps. Cody's left hamstring

is severely bruised, so I hold carefully as we lift his stumps
and place fresh chux pads underneath his stumps.

We roll the ace wraps over, under and around his stumps,
then give Cody a moment to lower his stumps

while we prepare his clean linen, then log roll him on his left side,
IV lines and all—I am now holding him

at the hip with my left hand, using my right open palm
up to my elbow as a bridge to hold his weight from falling back.

Stevens rolls off the dirty linen, unrolls the clean linen,
and Aguillera washes Cody's back with a soapy wash cloth.

"That feels good," Cody says in exasperation.
"That's what we're here for, Cody, to help you feel better," I reply.
The female nurse is angered—I turned the monitor off
and removed the electrode attachments for his ECG read,

though he is at peace—"That ringing was driving me crazy."

With three corpsman caring for him, we would know if he coded.

As we complete our work, his words wrap over, under, and around my mind: "I lost my wedding ring, my wife is gonna kill me."

TALIBRATZ

I take credit for coining a new term in the war—
Talibratz: Afghan children under the age of 13
involved in or suspected of involvement in
the planting of IEDs—usually helping
one of their parents or an adult, or under
adult guidance to plant said improvised explosive device.

Notify Webster.

Talibratz often pass through the ICU—never see
parents by bedsides, or hear of their whereabouts.
Guilty blast marks pepper-dot blood-black-pits into upper chest,
missing or ripped-up hands and arms, both legs wrapped with gauze,
patches cover blast wounds in the inner thighs and crotch—
the injuries of one crouching while planting an IED.

It's hard to sympathize with such children.

TERRORIST

Many junior enlisted speak this word—
a substitute for "detainee" or Taliban,
while "*Haji*" is amusing for inaccuracy—
carried over from a dirtier war south.

Maybe a youthful surge of testosterone
injects such color into language as
suspects enter with large black blindfolds,
our rank and name tape covered for security.

You can't help but feel a bit of spite when
you've seen so many young Army amputees
and Stars and Stripes-draped bodies
broken, motionless in the back of our Humvees.

BED CARE ON A DETAINEE

The detainee slowly prays
to Allah
as the pain in his stump may
be increasing—
what remains
of his left leg.
HM2 Gambino and I
reinforce
the bled-through gauze,
HM2 places
the abdominal pads
around the site,
covering crimson
as I hold each on,
and he begins to wrap
the ace wraps around
the stump to secure.
The detainee winces,
lifts his stump,
trying to move against
the pain.
I pull out drenched
chux pads, quickly
toss them onto the floor,
since a biohazard can
is not within reach, slide
fresh ones under his wound.
HM2 leaves to man
the night check-in
at Patient Admin.

I enter the sanitation room
to prepare wash supplies.

The detainee is a young guy,
beard caked with mud,
a fine dirt powders him—
dirtiest patient
I've bathed so far—
must have been the 50 cal.
shell that knocked him
on his ass,
flying into the air
and meeting the earth,
no glory as Icarus,
he lives,
his left leg ripped off
below the knee.
I wash him with warm water
out of a metal basin,
shoot in
a bit of shaving cream—
a little trick
Lt. JG Karp taught me—

supposedly
gets the patient smelling
crisp and fresh.
Water turns brown
as I work through the beard,
getting out those clods
of dried dirt,
do a quick wipe
of his face,
arms, working around the restraints
and the IV site,
chest, working around the leads,
then a fresh, new basin
with warm water,
some shower gel
this time.

Now to work on that hair.
I place a pink
port-a-potty behind his head
as a water catch, a towel
around his neck and shoulders
to absorb and comfort.
I lather his thick, short hair,
as most locals have great beards
and short, well-kempt head hair.
It's a process,
though I'm cleaning him for surgery,
as he will get at least one more wash
of his amputation site
to scrape away dead meat,
ensure nothing foreign
is in the wound site.

I've gotten most of the obvious dirt,
so I head to the sanitation room
again,
empty the basin,
wash out the silt at the bottom,
fill it again with warm water,
throw in 2 Betadine scrubs
to really clean—kill bacteria—
get the water frothy
with the pink liquid soap
dripping out of the small,
silver Betadine packages,
let every drop fall in,
throw out the old washcloth,
drying towels,
draw fresh cloth to wash
and towels to dry—this basin
for the whole body—
his arms, again, chest, again,
his groin and right leg.
He shivers from the cold—
I wipe the wetness off,

ask the Army guards to remove
the restraints so I can get
at the dirt better.

Engaging them in conversation,
I get some of the details
of the detainee's story—
caught planting an IED
with a friend, "who wasn't so lucky,"
as the short Mexican Specialist
puts it. His end I would like clarified,
for curiosity and a good story,
but I let it go.
The detached and distant know
that most of these guys plant IEDs
for the money, no religion
or politics involved,
as this is a piss poor country
where money and resources
are few, and shared only
through clans.

I work on his foot,
the Betadine scrub's hard,
plastic bristles sprinkle
dirty water on me.
I run the wet hand towel
between his toes,
wipe the brown foam
from his sole,
getting it all off somehow.
The nurse preps the bedroll
that is to replace the one
covered in dirt and water
from the bath I've given him.
We log roll him on his side
toward me,
a pillow between stump
and leg
to prevent painful contact
between the two—
arterial lines, IV lines, EKG leads
are all carefully gathered
to make sure they don't snag or pull out
with the old bed sheets,
or are mistakenly placed
under the elastic of the new bedding.
As the nurse wipes his back and ass,
I pull away the old sheets
and throw them on the floor quickly.
We lay him back down,
then log roll him toward her.
I get the fitted sheet corners in,
we lay him back down
and the bathing and linen change
are complete.

THE MARINE

A few nights ago I bathed a local/shot while planting an IED. Last night
I helped with the care of a young Marine/who was blown up by an IED

—journal entry

I'd give all for Commander Riser—
he's a living guide within a ghost ship
of passing souls—his dark humor and off
comments draw glances from other nurses,
timbre chuckles from somewhere inside me.

This night will be a pulling tug on us—
Bed 8 presents us with a young Marine.
The Respiratory Tech allows me
to pull the endotracheal tube clean.
A mess of blood-mucus follows it out

the passageway of the Marine—he coughs
with its movement, and I help clean and remove
the chux that has caught the blood-mucus mess;
He now breathes without a machine.
CDR Riser follows protocol,

questions the Marine; woozy-waking from
anesthesia, he gains wits and bearing.
A bilateral amp rests next to him—
a drawn curtain, each only observes
his own horror—neither quite aware.

"Why is there a pain in my side?" he asks to sky.
"You have a tube going in," comes CDR's reply.
"Ooohhh, but of course!" he responds, a humor
toward something unnatural, at this time
a part of life (if only for a little while).

He stabilizes: increased cognition,
strong vitals, able to breath on his own,
aware of pain not related to wound—
like that tube draining fluid from his lungs.
He wanders light-headed from anesthesia.

I change his set up due to status change
and an officer's medical orders.
First I remove the feeding tube from his nose,
in goes a nasal cannula—O2 levels sat—
now he can communicate proper:

"I want to see my sons, I miss them so much."
Two boys back home—both can barely speak,
tears run, he mentions his love for them
through the night—we work to keep him alive,
while the night burdens, from long to longer.

As a corpsman, I can only do small acts—

monitor vitals, help move him and give
him water, CDR Riser requests
antibiotics, and the show goes on,
all comedy dies with awareness:

"Boy, is my wife gonna be pissed!" Funny,
you hear that so often from the amputees
when they awaken from deliverance—
explosion/shock/helo ride/trauma bay
/OR/ICU. That worry of what

A female close to them will think of what
their body has become through war, as if
they forgot to buy groceries or walk the dog.
It's often the first thing they mention, concern
for anger from the wife. Sometimes it's Mom.

His awareness shifts from loved ones to self—
his condition, his current dilemma, his fate.
"Taliban! Bastards got me! Got me good!"
Yeah, looks like it was a strong toe popper—
turned his right leg into an odd-shaped wick.

Slow tears come, no comments now, on his own,
inside himself. We work around his stare.
Politely, CDR gauges the air.
I'm requested to help bathe and redress
the bilateral amputee. Who knew

it could be such a nice break—he can't talk,
so I work—not to listen, not to think,
free from conversation with the wounded,
painstaking task working on the body—
no mind. His nurse and I work mindfully.

Whenever 0645 comes around,
having worked through this longest of nights, up
to its end with my nurse, I give my hand
to the Marine—"Have a safe trip home.
It's been an honor taking care of you."

I never saw him again.

COAGULATED BLOOD

CDR Riser tells me to clean her up. She's an Afghan National—
as we call them—40 or 50 yoa, a coif of gray in her black hair.
I return from the linen stow with a washcloth, gloved up for BSI,
lean into her—slowly work the dried blood off her left cheek,
right cheek, forehead, being careful around the loose flesh—
flaps of skin on the upper right and left sides of her face,
her flesh is delicately held together and to her head by stitches—

as the meat was torn—almost ripped off completely—in an MVA.
A neck brace bejewels her, the stem of a blackening crimson rose—
something one doesn't always see, but she has a fractured atlas—
the circular bone which the skull floats upon, connected to the axis
by four complex joints collectively known as the atlanto-axial joint.
The ICU staff's greatest worry is that she may awaken in a panic,
thrash around in delirium and inadvertently break her neck.

Betadine and water moisten the coagulated blood coating her face
like a mask, glued into her hair, which is pasted to scalp.
The metallic odor of platelets comes back to life with the cleaning,
attacking my sense of smell. I silently gag my way through the task,
and curse my great sense of smell, though my face is unmoved.
I imagine vomiting, which would not bide well in any healthcare
setting, let alone a military hospital in a war zone.

The tear along the left side of her face is open—no stitching,
so the flesh is really loose. I place a wet cloth over it
to moisten the dried blood, and then work away on the right side,
cleaning out her ear to clarify whether or not the bleeding
is run-off from facial wounds or cranial leakage.
Nothing is coming out of her ear, and there are no signs of yellow
in the seepage, so she's safe, for someone with a fractured atlas.

Her face becomes clearer with each wipe, fine features keen to eyes—
a working thought—*she must have been something when she was young.*

THE LITTLE GIRL WITH A BULLET IN HER HEAD

She's about 3 yoa,
head well-dressed,
hands wrapped in gauze
and well-taped—
as she's been trying
to pull out her cath.
She fidgets,
her well-taped hands
rubbing together
like kitten paws
or cute cotton swabs,
while she chews
on her bottom lip,
her straight little teeth
pink with blood.

Though painful to watch,
she isn't cognizant.
Her x-ray shows a bullet
trapped in her frontal lobe,
"removing it
would be fatal,"
the floor doctor
kindly tells me.
She's listed
on 'comfort care,'
and moved to the Children's Ward
to free up
the ICU nurses
for patients
who might make it.

I pass through
the Children's Ward
three days later
wanting to know
the status of one
of the cutest
little patients
we've had at Role 3,
and find her bed
already empty.

ABDOMINAL PADS

I imagine them oversized
maxi pads
made for the goddesses
up high
on Mount Olympus—
how they absorb
as we absorb
as earth absorbs
those we cannot save.

A DREAM

JOURNAL ENTRY:

Had my first dream related to this place last night. It was so clear, it almost got filed away as a real event. I was secondary litter bearer, and a child was being bagged. I took over for the pj, and followed awkwardly up into the first Humvee, bagging the child. The flight line leader took over, I crawled through the hatch to the passenger seat, then got out and drove off in the secondary vehicle. Come to think of it...

It wasn't a dream...

NO MATTER WHAT

Resting on a patio
with a morning breeze
and it's Afghanistan,
eating a bagel
with cream cheese
and it's Afghanistan,
reading Orwell
among the Burmese
and it's Afghanistan,
Jack London
in the South Seas,
and it's Afghanistan.

SUDDENLY REMINDED

I forgot about the superficial
and mundane back home,
until I'm internet browsing
in my dorm room at NATO barracks—
see a civilian friend's
post on some social wall:
"Someone cheer me up. had a rotten day."
Wit leaves me replying:
"You're not in Afghanistan with me—
that should cheer anyone up,
unless they're in Afghanistan..."
to which
one of his young female friends
back in Southern California
replies:
"That's actually pretty depressing."

MOMENTS WITH A FALLEN SOLDIER

In Kandahar darkness, waiting off of the flight line,
not even the smoke deck lights are on any longer.
Deep into the night, corpsmen are crashed out in Humvees,
sleeping gaped mouth where wounded and the dead often rest.
This hot black well I'm dwelling in, waiting off Foxtrot,
I live in these moments, in desert emptiness,
green LED outlines from coms and those airfield lights—
red, white and blue—could never figure out their purpose,
though pretty to view—I feel peace—restful as a tired child.

We have an inbound Angel, a young soldier needs us.
Not our role—mortuary affairs is MIA.
We can't say 'no' to a soldier who needs to be brought
back home—we're but one stop on his return to loved-ones.
Blackhawks arrive, storm of wind and invisible dust,
in darkness the exchange occurs—soldiers to sailors
we carry their brother back to the Humvee I drive,
solemnly slide his litter into compartment rear.
Flight line leader sits up front with me, we make our way.

Down a dirt road behind the hospital, pull into
mortuary affairs—never knew it was so close.
Under yellow overhanging lights—a strange Christmas,
a dead fiesta, for those we couldn't save arrive here—
they aren't empty boxes, make no analogies here.
A small dirt courtyard, a few buildings—no signs of life.
We park, get out, HN Stevens looks for anyone,
I walk and observe under dim light, but don't wander—
the soldier's body is my responsibility.

I look inside the Humvee rear, his form vague, hidden—
Stars and Stripes his shroud, his dusty combat boots revealed,
general issue tan boots where red and white field ends.
Left leg points 45 degrees as one at rest would,
the other grotesquely inward—flat—the leg torn off
then placed back with body, or twisted with force of blast.
It offers the violence of his death, that awkward boot,
it forces me to feel for him most—inglorious
the body's destruction, but the folks back home won't know.

They'll only see his face—if that's intact—not about
to check, it wouldn't be right, I don't want to know how deep
the damage goes, Old Glory covers him, protects me.

In darkness away from fiesta lights, in darkness
I had long moments with him, what I said—between us—
no one else need know—though they were words of kindness, words
of solace, words of hope. It must have been all for me,
for in the cavernous hot night, I felt despondent.
All I could muster—the little I could imagine.

DEATH CARE

Stepping into the ICU, an officer asks me,
"Have you ever done death care?" I take on a new task,
figuring it's something that can help build my resolve
and increase my knowledge of medical processes.
There he is—slightly elevated bed, head facing
Mecca, ashen toes pointing towards me—an afterthought
to eternity, begging the question already
set as an order. I go to work, shut his eyelids,
pull intubation tube from mouth, his tongue warped and gray,
overcome with the lack of color death offers.

I pull leads off of chest, remove IVs from cold arms,
deflate foley cath balloon, remove it from penis,
wipe the body with a soapy towel until it's clean.
Lacking knowledge of Muslim rituals—I find Chris,
the Afghan interpreter in the Inpatient Ward—
ask for help. Chris brings in a local patient's escort.
While they're giving him proper care, I watch silently:
Close the mouth, bind it shut with a cloth wrapped around head,
bind thumbs together, bind toes together, cover him
with blankets so he may keep warm on the other side.

I watch silently,
the clinical coldness in me broken,
somehow touched by their tears
for a man they have never known.
I bow my head slightly,
maybe at the weight—
death is death,
maybe at the guilt—
this is what our presence helps bring,
maybe at this death—
humbling me toward my own.

A NOTE FROM THE MOON

I am fully engaged and a part of the moonscape now,
so used to it I almost lose all appetite, becoming
as dry and barren—to have my mind surrounded
by mountains, which themselves are dry and barren,
holding the watchtowers of ancient glory, in the sky tools
of modern glory's observation—pale blue gives way
to lifeless dust, gray and brown, which occasionally cools
the air; but this heat forces itself on me, contains me
like an insect stored for a spider's sustenance, age
sweeps over me in this quiet death and the immensity of nature
beyond the barbed wire. I cannot wait to get home
and find new heat, laugh a deeper laugh at ridiculous
green grass and blindfolded-payday-restaurant-goers,
imagine killing my large, stupid neighbor with my bare hands,
to be let go by the police on the Afghanistan wild card.
My glory the glory of passive assistance, pure understanding
when reading *Tropic of Cancer*, idealizing a future—
in a more living heat—the heat of the tropics,
wishing I had brought some Camus to take the sting off—just a bit,
waiting to be relieved of this dry barrenness
that has taken root in my mind and soul—wondering
if it is some *tabula rasa*, a divine trepanning to the next state
of pure being, of less awareness, of a Communist mind set—
this soulless landscape a physical actualization of the red book.
So what is next on the other side? What will home be like?

ANOTHER NIGHT IN THE ICU WARD

JOURNAL ENTRY:

Worked in ICU last night, as usual. Had a double and a triple amputee come in from the same incident. They were both well bandaged when I got in there. The triple was a big fat Army guy, kept shaking his head "no"—to what, one can only guess. Both were given Purple Hearts—like that means anything at this point. Busy mainly helping clean locals for Lt. JG Karp, Lt. Flynn, and Lt. Radcliff.

THE DOUBLE-AMP LIEUTENANT'S WIFE

"Ma'am, you may not want to be here for this."

There is a nervous laugh—my nervous laugh,
trying to downplay the honest intent of the statement.

He came in earlier in the night. I was right head on the litter carry.
He was reasonably big and broad—probably around 6'2" with legs,
but he was much lighter now, with nothing below the knees, though
still heavy enough to force the leg litter bearers to grunt and struggle
to lift him up high enough for the push into the back of the Humvee;

And there was the blood. Everyone on the litter crew got bloody.

Standing by in Trauma Bay in case they needed spare corpsmen,
we noticed it on our blouses and trousers—the helo's rotor wash
kicked it down my right leg—a long, dark stretch of chocolate on
my DCUs. The female HM2 was the worst—all down her trousers.
Carefully, I first pulled bloody latex off one hand, then the other.

9-line done, purple-blotched blue latex gloves gone, I hit the head,
washed the feel of it off of my hands, then wetted and soaped a towel,
working the length of the stain. The drops in my boots weren't going
to come out. The one and only time I got blood on my boots. Tonight
is a busy night, so no internet browsing, and I'm absent from ICU.

When all calls are clear, I finally make it over hours later. He's in Bed 6,
on a non-rebreather, which is good—he's able to breathe on his own.
The sheet is up to his clavicles, outlining his thick, strong body, the drop
of fabric where legs end. He has that waxy complexion of the severely
wounded. It makes them look like they're not real, like wax dummies,

like they're dead. He isn't conscious and his O2 sat fluctuates. It looks
like he'll make it, but anything can happen, and that's why our ICU
is one nurse, one patient. This one has to be watched. Not that he isn't.
He's an Army lieutenant, and it turns out his wife is as well. In fact,
she's on base, and in a chair next to his bed, hands gripping tissue.

Just stepping in, I am requested by the nurse to assist a young Air Force
Airman reinforce the stumps. Amputees bleed out post-op, the docs
washing out the injury, cutting away discolored tissue and jagged bone,
trying to even out the legs, if they can, for future prosthetic fitting ease,
finally closing it up, but leaving some room at the bottom, where blood

can ooze out, as it will. Gauze isn't too stained, I notice, and I'm happy
for her, because she's there watching, and the Air Force Airman is only
visiting our ICU, so he's a little nervous about the process, though
I'm calm, having done this enough times, knowing how to set everything
up for a quick wrap, the stump lifting, the gauze rolling, that ace wrap
final covering. And that's when I turn to her with my recommendation.
Easy as this process is, watching those stumps being lifted up, then
re-mummified... . I don't think it's something she would want to see.
A minute into beginning the process, she quickly gets up and walks out,
out of the ICU, not through the closest doors, but out the longer hallway
on the other side of the ICU. Maybe she's not thinking, or maybe

she wants to come into contact with as few people as possible. I try
not to think about any of it. I only think good, she's gone—that's better
for her, and it makes what we're doing less awkward for me, and less
nerve-wracking for the Air Force Airman. I alternate our tasks,
since he is learning. I wrap while he lifts, then I lift while he wraps.

That way he gets comfortable with the process. I'm not expert, but I'm
a pro from doing this numerous times now, on many Afghans, Americans,
English… I sympathize with the first wrap's strangeness,
that first questioning hold of a half-limb, and the worry of hurting the patient—the
greatest worry of all—that worry of hurting the patient.

DONUTS IN THE WAR ZONE

DURING A ROCKET ATTACK/TIM HORTONS WILL CLOSE.
IMMEDIATELY AFTER THE/ALARM.//WE WILL REOPEN APPOX. 15
MINUTES AFTER THE "ALL/CLEAR"//—thank-you//management

The line for blended mochas stretches out from the Canadian Compound
into the street leading to the weekend bazaar—a typical morning
on base—as caffeine brews to run troops who protect oil and lithium.

A Kandahar sun bearing on M16s, M14s, and M249s gun-barrel-hot,
and all of these soldiers—Canadians, US Army, Slovaks and Brits—
must feel like they're melting out of their uniforms as they wait

to get inside the Tim Horton's freezing air-conditioning, pick up
aforementioned blended mochas, a few glazes, sprinkles, chocolates,
and maybe a cheese bagel with the garlic and chive cream cheese

(my personal fave).

I'm in pt gear—Navy dark blue running shorts and bright yellow shirt,
my M9 strapped into my shoulder holster, and my wraparound shades
keeping me cool, and almost looking cool in military exercise gear.

It's my day off from hospital duties, so I'm hiding out from coworkers
in the Canadian Compound, and Tim Horton's—their version of Starbuck's
—is close enough to the hospital that I may see someone, though unlikely,

as most Americans won't walk over here, unless looking for privacy.
This is one of my refuges on base from the bullshit I deal with at work,
and I'm close enough that if there's a mass casualty alert, I can head

right over;

though right now, I'm just under the shade outside the recreation lounge,
John Fante's *Full of Life*, Larry Levis' *Selected Poems*, and my journal—
which helps me keep track of this alien landscape and all of the drama.

I might spend a lot of time on the journal today, as I saw enough shit
in the ICU last night—another young Army kid with missing limbs
and a purple heart hung on his pillow. I've seen enough purple hearts

to question the sanity of giving awards for wounds—albeit well-deserved,
but what do these awards mean to changed lives? Well…a purple heart
earns you lifetime commissary privileges, and gets your kids through school

free—now ain't that somethin'.

So I'll write it to down to remember, because working at Role 3
leaves too much to be remembered, and it's sometimes hard to revisit,
but I want to keep all of what I see clear in my recollections, for acquaintances,

those who are already forgotten back home by shoppers
and college students who don't need to worry about a draft this time around,
so they forget just as easy as well, and it's back to me, alone in my head

here in the warzone, enjoying a coffee and bagel, contemplating a double chocolate donut, hoping someday I might do justice to the moving pictures in my mind that replay often times; at times stills—frozen images;

but for now, I think I'll just have a donut.

CHOW CONVERSATION

Sat down for morning chow,
a long night shift over, the stumble
down the dirt roads of KAF
as it just starts to heat up at 0700.
The base quiet, the dust laid,
past the Boardwalk to the National DFAC
where I see HM1 Mursky—
a buddy who works in our blood lab,
and he actually invites me
to sit down—me, the most unapproachable
enlisted service member at Role 3,
but we go back to indoc
at Port Hueneme—where this all
started for us.

"Last, you missed it yesterday,"
he says in his nasal,
always-trying-to-find-the-humor
in-stupid-military-bullshit-way,
and he just loves the way
I roll my eyes
whenever policy and procedure
pile up to embrace the suck,
or whatever new events
at the hospital defy logic,
or are just completely insane.
"Why, what's up?" I raise
from my monotone,
to accentuate my eternally sardonic voice.

"We had a really bad call,
some Marines came in with IED
injuries. This one had his legs
and an arm blown off.
The doc on duty asks him,
'Are you ok?' and he says,
'What do you mean,
am I ok?? I only have one arm!' "
There's a chuckle,
and I chuckle. It's that darkness—
the brute honesty that attacks
a question that is truly
insane, yet a requirement
in order to process patient cognition.

It's when protocol goes
out the window,
when an enlisted can basically
tell an officer to "FUCK OFF!"
and fuck off with righteousness,
knowing nothing can be lost.
It's the cruel ridiculousness
of what we see almost every day
coming into the hospital,

or residing in it until death
or some almost happy ending
do we part with patient,
though where they come from
tells a lot about their chances.

In the case of locals,
some have a slow death in our ICU.
In the case of our troops,
often a long flight to Lundstuhl,
followed by a longer flight
home… a bed at Walter Reed
or Bethesda down the street.
But me and HM1 Mursky,
we can only chuckle
at all of the insane bullshit—
we've been here months
and have months to go.
If we can't laugh at cruel fates
we'll become victims as well.

UNCONSCIOUS

I laugh at the futility of the female nurse
talking to the unconscious Afghan male.
True, people are often aware, even while
sedated, but the care in her tone attacks
my reality, seems so stupid. He is unable
to grasp consciousness, but she goes on.

TO DEATH

Marine pilots rage overhead,
the obscene sound of sky tearing
as they buzz the base, while I'm
tanning by the barracks wall.

I can feel the swelling
in their testicles from down here—
so can my ears.

Maybe they do it for the rush
or the immediacy of fire support.

I am left deaf, distracted, wondering
the reasoning behind this assault of noise,

demystified by the joy in war,
for their joy in giving death,
as if it were a present for their own child,

and my occasional love for it as well.

SANGFROID

HN Hightower is blowing his lid,
red-faced-yelling over my beanie-wearing
indoors, while I'm laughing and pissed at once.

      "HM3 LAST, REMOVE THAT BEANIE—
        THAT IS A DIRECT ORDER!!"

—and moments later—

      "TAKE THAT BEANIE OFF,
        YOU OLD BASTARD!!"

Final words of a young, hard-wired jackass;
at my cubby geared up for a 9-Line,
I turn to tell him, in a moderate voice:
"It's a bit more unprofessional for you
to be cussing at me than for me to wear
a beanie in this ass-cold hospital."

My response surprises me—
how I keep composure, I don't know.
9-Line complete, I'm at smoke deck hanging
with my buddy from the Inpatient Ward,
HM3 Dean, bitching about that asshole.
Our chaplain slowly draws from his pipe, begins:

"He's had seven patients come in dead,
and wasn't able to bring one back.
Keep in mind that these active duty kids
are going back to Navy clinics where
they won't be appreciated, where
sailors fake sick to get out of work.
It's going to be a hard adjustment."

An officer and advisor, Chaps gets my nod,
a sign of respect on all levels.
"The mission is first," Chaps reminds me.
I nod in agreement, though logic
leaves me thinking in the darkness, through his smoke:

but I'm part of that mission, and if
I'm distracted from carrying out my duties
by the shortfalls and ignorance of other people,
isn't the mission threatened?

FALSE ALARM

The fire alarm goes off in Cambridge DFAC—
I just sat down to dinner and am pissed.
I know it's a fire alarm, not for rockets.
A fire alarm wouldn't go off if we were hit,
you'd have that English woman's proper voice
sounding off, over and over: "*ROCKET ATTACK...*

ROCKET ATTACK," but these dummies
jump underneath tables, lie flat to ground.
Finally, I'm the only one eating.
I get up, shaking my head, slowly push
out my chair—join stupidity—
as this is the military—if you
stand out, someone is going to give you shit.

After a minute, soldiers are pulling
themselves up, laughing, realizing what I knew.
I get up, wipe the front of my blouse,
go back to eating my meal, shaking my head,
wondering how much longer I have
to deal with rockets and ignorance.

A CIGARETTE AND A MARGUERITA

Jo Allahandra, as was his patient name made up in PAD,
fell off a horse, landed on his head, and is now in Ward1
of the Inpatient Ward, a hole the size of a golf ball drilled
into his skull to alleviate pressure from brain swelling.

When he first came in he was tied up to his bed rails,
all four limbs strapped down so he couldn't thrash
as needles were put into his arms, while altered mind
struggled inside a bed-bound body, wishing to move.

His bowels were out of control as well, but couldn't be tied,
as he shat himself constantly throughout shifts—day, night,
and we found out as corpsmen what great nurses we had,
as they were willing to clean up before we could get to him.

One night coming in, the nurses were talking about a female
officer who had completed a shift with six bed changes,
laughing about the doctor's comment when viewing her face—
"She looks like she could use a cigarette and a marguerita."

My first shift under a Canadian male nurse, we cleaned Jo
down three times, changed his bedding each time due to mess,
and once, as his ass was exposed for wiping, he let out a charge
of gas, as if to say, *Fuck you, thanks, and here's more to wipe.*

And so Ward 1 became the least desired ward for corpsman
during Allahandra's stay, and as his father sat by his side
throughout the days and nights, he must have been impressed
by the Americans' patient care and habits toward his son,

though to a young corpsman friend, Bed 2 earned the nickname
"Sir Shits-a-lot", and I would laugh if I didn't have the memory
of smells from his wet liquid messes and constant scrambling
that made blood draws and line changes long, arduous tasks.

WAKING REFLECTION

Waking from a dreamless night

afternoon light breaks

the blanket curtain

and I need some light to get me through

the long night shifts

under fluorescent light

glowing back from cool rainbow splatter-

pattern hospital floors.

Cream walls neuter

such brightness, though the wounds

treated are clear under light,

clear as a dreamless night.

TANNING IN AFGHANISTAN

This olive green cot allows me some rest as jets scream,
helos thump, and Black Sabbath plays on my iPod player—

found said cot in a rock and gravel pocket of NATO barracks
—no one's name on it—took it to my room, wiped it off.

Now I lay out on it under the afternoon Kandahar sun,
an Afghan prayer blanket and a beach towel my cushion,

they collect the sweat off my body as I tan in an alley
along the furthest brick wall, against green tarp fence,

barbed wire with captured-dirt-covered-plastic my view,
the heat and smell of Coppertone take me somewhere—

not Guam, not Hawaii, not Mexico, not even California—
nothing to compare the dry heat here to—nor the dust—

but I need melting penetration—waking at 1300 after another night
in the hospital, burning trash pumping through the a/c,

the lights so bright everything was obvious, though in darkness
Ward 3 only commands scents of sleepers—copper/iron incense

of blood and sweetness of wounds—some will never heal;
that darkness at the nursing table of the Afghan National Ward—

the sun prepares me, not in a literary sense, but within senses—
if I can bear this heat, let it penetrate flesh, puncture lungs,

I can go on treating patients felled by blasts of heat—no metaphors.
I turn up the music, laugh with the Slav soldiers at a bench

in their short shorts cleaning black steel rifles and heavy weapons,
working on mean sunburns, safely exposed as we all might be.

A ROOM OF CRIES AND PAIN

They cry out from the darkness of that room.
Even when the lights are on,
it stays a house of pain, a hurt locker,
whatever you want your civilian mind frame
to poetically conjure;
but the young enlisted guys are DnD geeks
and gamers, and the Afghans
are yelling in pain, "*Darkee!*"
to get our attention, to get pain meds,
to grab for something;
and so Ward 3 has earned the dark-humored nickname
'The *Darkee* Dungeon.'

You just look at our board:
You'll see that Bed 18 has TBI and facial injuries.
Bed 19 next to him is a GSW ABD—
I've seen guys in ICU die after two weeks
of struggling on a respirator
with the same wound.
Bed 20 is a globe injury—eyes damaged
from an IED blast.
I've had to administer quarterly
eye drops to him
and change his eye shields and dressing.
Bed 22 has a head injury
and shrapnel in his back and has been
a pain in the ass—we've had to strap him down,
since he's been trying to roll out of bed
and pull out his cath.
Bed 23 is a GSW/L Arm
(I almost want to yawn at that).

These are the current dwellers
of the dungeon, and the sympathy is weak,
as the cries of pain stab deep,
and young corpsmen grow callous
in the everyday of blood and trauma.
Bed 20 is a young Afghan National Policeman
with deep facial blast wounds.
After washing him, I open the little purple
Bacitracin packets, putting ointment
on my blue latex gloves,
and gently cover his skin.
Topical antibiotics aren't the best thing,
as some wounds are filled with packing tape;
but it's more than what most are doing
for him, and I feel for a 19-year-old kid
whose gray-misted eyes will
most likely never see clearly,
if at all. I administer his eye drops,
and am able to grab the *tajiman*
to help explain to him the process,
as he's always yelling for the *tajiman*
in panic, as a lot of them do—

crying for pain meds, moaning aloud
"Darkee"... "Darkee"... "Darkee"...

A CHILD WITH BLAST INJURIES

JOURNAL ENTRY:

Had a kid last night in Ward 1 with Fox eye guards, a missing right leg, jacked up left leg, fucked up right arm, and he was still able to swing his chux-covered arm around—made getting blood difficult—Lt. Radcliff ended up drawing—a forty-five-minute spectacle with four different nurses trying their luck. Currently at the British compound enjoying the a/c and *Borat* on the *telly*.

GONE

One guard possesses the keys
to the brown leather restraints
that securing him demands—
always two: one on a wrist,
the other on a leg, or both legs
or both wrists, as in one case—
a *Haji*—sole survivor
of a truckload of armed men
who met the steel weaponry
of an Apache helicopter.

I remember the first night
I cared for him in Bed 30,
in the side room for detainees.
He had hate-filled eyes and both legs.
The next day he went under the knife.
The next night, on my return,
his outline...his left leg—gone—
amputated above the knee;
from right leg rose sweet and sour
smells of decay within days.

Some nights later, I return,
his form through sheets, right leg—gone—
a bilateral now, eyes passive,
their fierceness—gone—surrender
to restraints, the death stare—gone.
Never before spooked by a *Haji*,
until his first glare at me.
Maybe, with both arms restrained
and both legs gone he realizes
I might be the last to care for him.

THE INDIFFERENCE

This room, where
allies become Taliban,
where the care is bitter,
languages pass one another,
turning an ally, a sympathizer,
into the hate-filled, jihad-obsessed;
the silent broken moments in this room,
when the interpreters are on a smoke break,
or just not present to explain to those bedridden
why one of their own cries in pain as we move him.

We need to clean him, rotate him to prevent bedsores,
prevent him from hurting himself any further,
administer treatment to help him heal—
it will hurt like hell while performed;
but they only hear the loud cries,
only see us struggle with him,
see him struggle with us,
crying out to Allah,
to the *tajiman*,
to them.

SOMETHING TO PONDER

I don't think any kind of freedom
for the Afghan people
from the Taliban,
Iran, Al Qaeda, or even Russia,
is worth the death and damage
done to young Americans;
and quite often
there is not a look of thanks
in the eyes of the bed-ridden
Local Nationals in our ward,
but looks suggesting,
"Why don't you help those suffering?"
"Why did your people do this to me?"
"Why are you here?"

Good questions, dudes...

SOME PEOPLE JUST DON'T GET IT

We're stacked up off the flight line
and I'm with the new HM3 who's part
of the relief for the thirty who arrived
here a month before our glorious group,

and he says to me in an even tone,
straight-faced, "I don't even consider
this a real deployment." I just look
at him quietly, gauging for humor;

but I don't even sense dry sarcasm,
as my mind engages—trying to find
the logic in a statement as absurd
as this ridiculous war. It fits right in.

He's at the end of seven years' service,
and maybe he's used to being on ships,
or at a command by the water where he
took care of the 'blue side' sailors—

they treat their corpsmen like shit;
or maybe he thinks being stuck on a base
in a war zone isn't really being
in a war zone—as if the heat, dust,

continual exposure to run over, blown up,
burned up, shot up, cut and fallen
human beings just ain't the glory
that it should be, not like an 8404—

out with the Marines receiving direct fire,
the sky for your blanket, a rock for your pillow,
as the ridiculous romantic cliché goes.
Maybe the NATO dorms are just too comfortable.

I let it go. I want to ask him how much
he's seen so far, maybe that's what it is.
I let it go. His ignorance is his bliss, right?
Hopefully he doesn't wake up one morning

realizing exactly how wrong his statement is.

COLOR

Bland tan, dulled silver and gray—all around me.
Barbed wire rings, concrete walls, rocks, gravel
and fine dirt powder. The only color:
uniforms, red brick of NATO barracks,
and wild canaries at the Australian compound.
I try to forget the cast that blood, organs,
and decayed tissue endlessly offer—
chance it could be me, then I am my own art.

I write in color—red ink of my pen
commits memory to page. I write possessed.
I want to remember what I've seen, and
if I can't, to record for those who cannot,
whether in passing, or for lack of proper words.
If these words are only mine, so be it.
I trust them, even if they are not the right ones.
Their sincerity is all I have—I bring color here.

REFLECTION

JOURNAL ENTRY:

I feel like I have been compressed in a bottle and have floated for a near-eternity,
like a being about to be released, having washed upon a mysterious beach.
Sitting at Tim Horton's in the Canadian Compound brings this to me. I don't quite
get it. There are a few situations and places at the hospital that give me this
feeling as well—the walk to smoke deck under darkness, the bright-lit ICU.

It's strange to think that this hot, dusty stasis is being removed from me like
a meal at a table, and something else put in its place.

HOME
01SEP2011 - PRESENT

WALKING THROUGH LAX

Walking through LAX in my ACUs,
I feel like a freak show.
Every other airport our group has passed through,
we get at least one "thank you."

Here, in the land of Hollywood,
I feel silent, judgmental eyes observe us,
or ignore us—we're suspect—

after having watched too many conspiracy films
where those in the armed services are faceless
automatons who do their government's bidding.

I have more individuality in my proverbial
little pinky than most of these assholes
have shown in their entire lifetime—

and that would include the Chinese tourists
waiting out by the curb.

BACK AT NOSC LOS ANGELES

I report back to my Navy Operational Support Center,
the reserve station where I drill one weekend a month,
wearing my service khakis, or peanut-butters, as is the slang—
since it's a black and tan outfit—black pants and an off-brown
button-up shirt, and black service cap with rank designation.
I call it the Starbuck's worker's outfit, since I feel like I should
be serving coffee every time I wear it, and knowing the Navy,
that's exactly how they want the enlisted ranks to feel.

I want to wear all of my ribbons though, two new glorious rows—
never gave a damn about the politics, but these people do,
so I'll wear my three rows of medals to shut them up, so that they know
"a fucken Reservist" did something in his Navy career.
I patiently wait to sign paperwork to be squared away,
the PS1, in casual conversation, can't be bothered
and a PS2 I get along with enquires where I was deployed.
When I tell him "Afghanistan," he responds, "Oh, so it wasn't that bad."

Our clinic's HM1—military bearing in need of polish—
has a tendency to bark down to those she outranks—
walks by me like I'm not there; same said HM1, when I asked her
what I would be doing in Afghanistan, responded,
"You'll be patrolling with Marines" (not true, I was to find out later—
you have to be 8404). She could have been clueless,
but given her history of storming into the clinic
hollering, "HM3 LAST!!" there's no love lost between us.

The CO requests me in his office, shakes my hand, as well as
the XO—a Navy Seal Lieutenant, who when shaking my hand
as I left on deployment, looked me square in the eyes and said,
"Kick some ass in Afghanistan." It's good to see him again, his polite,
determined look, as he squeezes my hand in a strong handshake.
One of the medical officers, back from Kuwait, is by the check-in
as I'm getting ready to walk out, having officially reported back in.
"Hey, you guys had it tough over there. I've heard. Good to have you back."

Yeah, it's good to be back—maybe not here—but back.

CORPSMAN DOWN

It's been two weeks at home,
and I manage to get downstairs
to the mailbox most days,

but I've been up in my room
and only out in the kitchen to cook,
have a tea, take a break

from the room, from the bed
I've been lying in watching tv, reading
little, waiting for the next thing.

I can't stay here much longer,
because it feels too good in this bed,
warm, safe, clean and familiar.

COFFEE SHOP CONVERSATION

AFGHANISTAN gets dropped
like a dead relative,
a bad pick-up line,
or an irrelevant interjection,
as each and every time
I talk to someone
and they ask me
the general questions
"Weren't you in Japan?"
"Weren't you on Guam?"
two places I continually haunt,
I say, in demurred surprise,
"AFGHANISTAN."
Then that strange silence
that is the norm.
It's not like you can follow
it up with anything, like,
"So, howse the weather
in Afghanistan this time of the year?"
"What were you doing over there?"
Or even the more benign,
"How was it over there?"
but... just a simple,
"How are our troops doing over there?"
or the sometimes cheesy,
"Thank you for your service"
would be better than that silence;
but this is Southern California,
where it seems everyone
reacts to any and everything
with niceties, and a word
like AFGHANISTAN makes
the robots malfunction,
penetrates the circuit board,
overloads the system:
snow-covered hills
dust covered plains,
fresh IEDs awaiting detonation,
training camps in Pakistan
awaiting Predator penetration,
and the coffee conversation
is cheery, benign,
even if you've been to
AFGHANISTAN.

"THE HUMAN TOLL"

This is the first time I have shed a tear over Afghanistan, and I am all the way
back home, tanned and reflective from a month in Hawaii, when I pull out
the *Navy Times* I bought at the convenience store at the Hale Koa Hotel.

Having a coffee at the Starbucks up the street from my place in Huntington Beach,
trying to ignore the shallow, high-pitched banter of the black-clad female workers,
espresso machines, loud customers in everyday conversation,

And I am finally finishing the paper I have overlooked. I get to the center,
where the war casualties appear. The column is entitled, "The Human Toll."
It's been following the same basic pattern for the past one-and-one-half years:

One hapless guy killed in Iraq, between 4-to-10 killed in Afghanistan, roughly
 around 20, 21, 22; occasionally one 30-something, slightly obese, IED victim;
and lots of first sergeants—always seems to be hit the most, that rank.

It's always Marines and Army—if it's a Marine, usually in Helmand Province,
Army in Kandahar or one of many hard-to-pronounce Northeastern Provinces.
Occasionally a corpsman out with the Marines gets hit and I pay my respects.

I won't leave out the most important piece of information—this is a weekly—
this goes on every week—it's the small ticker tape on the bottom of the screen
when you watch CNN or MSNBC or even Fox News. It's right there, ignored.

I see this young Army kid, killed by small-arms fire, Kandahar, Afghanistan.
I realize I had talked with him back at the Role 3. I can't remember if he was
a walk-in patient in Urgent Care, or guarding a detainee in the Ward,

but I remember his face, his last name, and he's 20-years-old, hit right before
Thanksgiving. This is what his family faced during the holidays, and here I am,
almost a month later pulling out an old paper to see him; my eyes glaze.

Nothing falls—there is heat in my face, moisture around the corners of my eyes,
and I think once again of the waste. Feeding our young to a country run
by warring clans. He died in the moon dust of Kandahar...

I didn't even know this kid, he just passed through the hospital, but it hits
something in me, reminding me of all those we have lost, the personal loss
of his family and friends back home. The life he no longer enjoys. All is gone.

I'm not crying. Getting out of the store, away from these young girls, the people
in this store who have no idea, who may know someone over there; but they don't
know "Over There." I'm walking back up to my apartment.

THE UNLABELED FILE FOLDER

I decide not to stick a book title
on the outside of the file folder
containing journals, pamphlets, field manuals,
Stars&Stripes, and random acts of paperwork
which will act as guides and memory enforcers
for my book on the war, lest,
while working on it at a coffee shop,
I get looks:

looks of pity,
looks of anger,
looks of understanding,
quizzical looks,
about-to-ask-me-a-question-looks,
I-could-give-a-fuck-that-you've-been-to-Afghanistan-looks.

I was weird enough to begin with,
I don't need anything to call attention to that fact.

SMOKING CIGARS AND DRINKING HOT SAKE

That's what I do on my balcony that overlooks the carport
of my apartment complex and the palm trees next door.
The blue and white-striped awning shields me from the sun
while I listen to some Rasta on my iPod speaker dock,
getting a slight, dizzy buzz from a *Partagas* 1945 Black Label
and some cheap, microwave-heated *Gekkeikan* sake.

Children scream in their wild play in the neighborhood
across from the boundary wall not quite high enough for respite,
and the car dealership way down the block blares its music
and announces great deals with low financing which even
a strong wind can't deflect, nor the barking dogs nor gardeners
in my neighborhood running lawnmowers and leaf blowers.

Over time I've learned to phase this all out—somehow I grew
from being over there, though I sit in my room, too comfortable,
wondering when I will break from this new womb and get out,
get it all out into the light to be seen. I sit working on the poems—
digging a little deeper into the hole—wonder over the deepness.
"Fuck, what am I doing combing and endlessly editing this shit?"

I mutter aloud, knowing this may not be helping, but it may
help. I write about my life—it is my work—and this work,
in particular, is most important, for there is a war that has
to be delivered in a clear, dirty, raw way that will hopefully
stir a passive audience and help them understand what our people
were exposed to—their sacrifices, their lives, this nightmare.

To take the unsettled self in hand, under whatever conditions,
and return to the mind with which one set out;
to pick oneself up again, after the mind changes,
weakens, and breaks down, and stiffen the determination;
to carry through the oft-reconstructed original vow—
isn't that the meaning of courage?

—Soko Roshi

GLOSSARY

ABD—abdomen/abdominal area

ACU—Army Combat Uniform

AHLTA—electronic medical record (EMR) system used by medical providers of the U.S. Department of Defense (DoD).

BSI—Body Substance Isolation. Protective equipment worn to prevent exposure to potentially dangerous body fluids.

CASEVAC—Casualty Evacuation. Interchangeable with MEDEVAC, but used at Role 3 as a term for the Air Force's wounded treatment and loading area.

Cath/foley cath—a flexible tube passed through the urethra and into the bladder to drain urine.

CDR—Commander. Naval officer ranking O-5.

CO—Commanding Officer

Comfort care—end-of-life care that helps or soothes a person who is dying.

Corpsman—Navy enlisted personnel trained and designated to give first aid and medical treatment. Also known as a Navy Corpsman or Hospital Corpsman.

Cranial—protective helmet used while on the flight line.

DCU—Desert Camouflage Uniform

DFAC—Dining Facility. Also known as a 'chow hall' or 'galley.'

Darkee—Afghan for 'pain.'

ECP2/ECP5—Entry Control Point. Heavily guarded base access areas.

EKG—Electrocardiogram. A test that measures heart activity.

EVOC—Emergency Vehicle Operations Course. Urgent Care Corpsmen were required to go through PowerPoint and hands-on training with the departing rotation, but the term was used as a misnomer, as personnel were assigned as EVOC1, EVOC2, etc.

Ex-fix—splints, plastic dressings, or more typically at Role 3 metal transfixion pins used to hold together fractured bones.

Foxtrot—Call sign for the airstrip next to the Role 3.

GSW—Gunshot Wound

Haji—a derogatory term for an Islamic fighter. Used in the Iraq Campaign and carried over into Afghanistan.

Heplock—a small tube inserted into a vein in order to allow quick circulatory system access for supplying blood products or medications to a patient.

HM (HN/HM3/HM2/HM1)—rate designator of a Hospital Corpsman. HN is an Enlisted E3, HM3 an Enlisted E4, etc.

Indoc—Indoctrination. Instruction related to a school, command, or rate. A term used for processing a service member when going on a deployment.

IV—Intravenous. Tubing system allowing easy access for blood or medications into the body via the circulatory system.

Lt. JG—Lieutenant Junior Grade, or Lieutenant JG (the more-common term). Naval Officer ranking O-2.

MA (MA3/MA2/MA1)—rate designator for a Master-at-Arms, the Navy's Enlisted personnel in charge of security and law enforcement. Same as the Army's Military Police/MPs.

MEDEVAC—emergency removal of sick or injured personnel by helicopter.

MOBBED—Mobilized. Sent on a specific mission under military orders.

NVGs—Night Vision Goggles

O2—Oxygen

PAD—Patient Admissions

PHA—Physical Health Assessment. All Reservist Sailors and Officers are required to have an annual health screening in order to determine deployment readiness.

PJ—pararescue specialist. A term typically used for Air Force CASEVAC personnel, but used at Role 3 for the Army MEDEVAC personnel we worked with—another misnomer.

PNA—pneumonia

Poo Pond—a water reclamation system on base that was basically a shit-brown lake that gave off an odor that matched its color.

PS1—Personnel Specialist First Class. PS is the rate designator for Navy personnel who handle most administrative duties and personnel records.
Role 3—a medical treatment facility (MTF) in a theater of combat that can supply total care, from dental needs to minor or traumatic wound and injury care.

Tajiman—Pashto for interpreter.

TBI—Tramatic Brain Injury. Degrees of injury vary, but at a minimum one has to have had their 'bell rung,' i.e. blacked out or experienced an altered level of consciousness due to a head injury caused by exposure to an explosion or as the result of an accident.

Tim Horton's—Canada's answer to Starbucks, with a heavy emphasis on donuts.

TOC—Tactical Operations Center. Used as an intermediary in communications between the MEDEVAC crews and the hospital staff—usually occupied by a lower-ranking officer or senior enlisted member.

XO—Executive Officer. The second-in-command at a military installation.

yoa—years of age

8404 (eighty-four-oh-four)—Field Medical Service Technician. A Navy Corpsman trained for 3 months in the combat care of Marines. An 8404 corpsman is the field medic for Marine combat units.

9-Line—a system of MEDEVAC designation indicating status of incoming wounded personnel. Typically Urgent Care corpsmen received something more akin to a 2-Line. Alphas were classified as litter-bound life-threatening traumas, Bravos as litter-bound intermediate traumas, and Charlies as ambulatory patients.

About the Author

Curt Last lives in San Clemente, California. He earned his Bachelor's Degree in Pre-Law from the University of California, Santa Barbara and his Master of Fine Arts in Poetry from California State University, Long Beach. He served from 2008 to 2016 as a Hospital Corpsman in the United States Naval Reserves. Duties included various Navy clinics and hospitals, a humanitarian mission to East Timor, and a deployment to the Role 3 Combat Hospital in Kandahar, Afghanistan.